SRA

Connecting Math Concepts

Level B Student Assessment Book

COMPREHENSIVE EDITION

A DIRECT INSTRUCTION PROGRAM

Mc Graw Hill Education

Bothell, WA • Chicago, IL • Columbus, OH • New York, NY

MHEonline.com

 Education

Copyright © 2012 The McGraw-Hill Companies, Inc.

All rights reserved. No part of this publication may be
reproduced or distributed in any form or by any means, or
stored in a database or retrieval system, without the prior
written consent of The McGraw-Hill Companies, Inc.,
including, but not limited to, network storage or
transmission, or broadcast for distance learning.

Permission is granted to reproduce the material contained
on pages 1, 3–4, 7–8, 11, 15–16, 21–22, 33–34, 39–40,
45–46, 51–52, 57–58, and 65–67 on the condition that such
material be reproduced only for classroom use; be provided
to students, teachers, or families without charge; and be
used solely in conjunction with *Connecting Math Concepts*.

Send all inquiries to:
McGraw-Hill Education
4400 Easton Commons
Columbus, OH 43219

ISBN: 978-0-02-103596-0
MHID: 0-02-103596-2

Printed in the United States of America.

20 21 LOV 26 25 24 23

The *McGraw-Hill* Companies

Mastery Test 1

Name _____

Copyright © The McGraw-Hill Companies, Inc. Permission is granted to reproduce for classroom use.

Part 1

a. _____   b. _____   c. _____   d. _____

Part 2

a. _____   b. _____   c. _____   d. _____

Part 3

a. $7 + 1$ d. $9 + 1$ g. $5 + 1$

b. $1 + 1$ e. $4 + 1$ h. $2 + 1$

c. $3 + 1$ f. $8 + 1$ i. $6 + 1$

Part 4

a. $30 + 10$ c. $60 + 10$

b. $80 + 10$ d. $20 + 10$

Independent Seatwork

$6 =$ _____ $3 =$ _____ $5 =$ _____

10   11   12   1__   __4   ___   16   ___

Remedies

Name _____

Part A

a. _____   b. _____   c. _____   d. _____

Part B

a. _____   b. _____   c. _____   d. _____

Part C

10 ___ 30 40 50 ___ 70 ___ ___

Part D

a. _____   b. _____   c. _____

Part E

a. _____   b. _____   c. _____

Part F

a. $9 + 1$ c. $5 + 1$ e. $2 + 1$

b. $6 + 1$ d. $8 + 1$ f. $3 + 1$

Part G

a. $60 + 10$ c. $80 + 10$

b. $20 + 10$ d. $30 + 10$

Copyright © The McGraw-Hill Companies, Inc.

Connecting Math Concepts

Mastery Test 2

Name _____

Part 1

a. _____ b. _____ c. _____ d. _____ e. _____

Part 2

a. _____ + ___ = 37   c. _____ + ___ = 52

b. _____ + ___ = 81   d. _____ + ___ = 43

Part 3

a. $2 + 7 = 9$

_____

b. $30 + 8 = 38$

_____

c. $15 + 1 = 16$

_____

Part 4

a. $2 + 1$
   $12 + 1$

b. $6 + 1$
   $16 + 1$

c. $3 + 1$
   $13 + 1$

d. $7 + 1$
   $17 + 1$

Part 5

a. $8 + 1$
   $8 + 2$

b. $3 + 1$
   $3 + 2$

c. $5 + 1$
   $5 + 2$

d. $7 + 1$
   $7 + 2$

Part 6

a. $9 - 1$ d. $8 - 1$ g. $10 - 1$

b. $6 - 1$ e. $2 - 1$ h. $5 - 1$

c. $3 - 1$ f. $4 - 1$ i. $7 - 1$

Copyright © The McGraw-Hill Companies, Inc. Permission is granted to reproduce for classroom use.

Mastery Test 2

Name _____

Part 7	Independent Seatwork	
a. $9 + 0$ $9 + 1$	a. $40 + 10$ $47 + 10$	e. $80 + 10$ $83 + 10$
b. $6 + 0$ $6 + 1$	b. $70 + 10$ $72 + 10$	f. $50 + 10$ $51 + 10$
c. $4 + 0$ $4 + 1$	c. $30 + 10$ $36 + 10$	g. $20 + 10$ $28 + 10$
d. $10 + 0$ $10 + 1$	d. $60 + 10$ $64 + 10$	h. $10 + 10$ $19 + 10$

14   15   ___   ___   18   ___   ___

10   20   30   ___   ___   60   ___

45   46   47   ___   ___   49   ___   ___   51   ___

$3 =$ ___ $6 =$ ___ $2 =$ ___ $5 =$ ___

Copyright © The McGraw-Hill Companies, Inc. Permission is granted to reproduce for classroom use.

Remedies

Name _____

Part A

a. _____ b. _____ c. _____ d. _____ e. _____ f. _____

Part B

a. _____ b. _____ c. _____ d. _____

e. _____ f. _____ g. _____

Part C

a. _____ b. _____ c. _____ d. _____

e. _____ f. _____ g. _____ h. _____

Part D

a. ____ + __ = 37   c. ____ + __ = 52

b. ____ + __ = 81   d. ____ + __ = 94

Part E

a. ____ + __ = 43   c. ____ + __ = 61

b. ____ + __ = 25   d. ____ + __ = 86

Part F

a. $2 + 5 = 7$ b. $14 + 0 = 14$ c. $15 + 1 = 16$

_____   _____   _____

Copyright © The McGraw-Hill Companies, Inc.

Remedies

Remedies CONTINUED

Name _____

Part G

a. $4 + 1$
$14 + 1$

b. $6 + 1$
$16 + 1$

c. $3 + 1$
$13 + 1$

d. $5 + 1$
$15 + 1$

e. $7 + 1$
$17 + 1$

f. $2 + 1$
$12 + 1$

Part H

a. $3 + 1$
$13 + 1$

b. $7 + 1$
$17 + 1$

c. $2 + 1$
$12 + 1$

d. $1 + 1$
$11 + 1$

e. $5 + 1$
$15 + 1$

Part I

a. $8 + 1$
$8 + 2$

b. $3 + 1$
$3 + 2$

c. $6 + 1$
$6 + 2$

Part J

a. $2 - 1$

b. $4 - 1$

c. $8 - 1$

d. $10 - 1$

e. $3 - 1$

f. $6 - 1$

g. $9 - 1$

h. $7 - 1$

i. $5 - 1$

Part K

a. $7 + 0$
$7 + 1$

b. $4 + 0$
$4 + 1$

c. $9 + 0$
$9 + 1$

Copyright © The McGraw-Hill Companies, Inc.

Mastery Test 3

Name _____

Copyright © The McGraw-Hill Companies, Inc. Permission is granted to reproduce for classroom use.

Part 1

a. $7 \quad 2 \rightarrow 9$

_____

_____

b. $5 \quad 1 \rightarrow 6$

_____

_____

Part 2

a. _____

b. _____

Part 3

a. $7 \quad 2 \rightarrow 9$

_____

_____

b. $5 \quad 1 \rightarrow 6$

_____

_____

Part 4

a. ____ + ____ = 90

b. ____ + ____ = 19

c. ____ + ____ = 15

d. ____ + ____ = 50

Part 5

a. $\begin{array}{r} 2 \\ +1 \\ \hline \end{array}$
b. $\begin{array}{r} 1 \\ +5 \\ \hline \end{array}$
c. $\begin{array}{r} 0 \\ +1 \\ \hline \end{array}$
d. $\begin{array}{r} 9 \\ +1 \\ \hline \end{array}$
e. $\begin{array}{r} 1 \\ +6 \\ \hline \end{array}$

f. $\begin{array}{r} 1 \\ +1 \\ \hline \end{array}$
g. $\begin{array}{r} 3 \\ +1 \\ \hline \end{array}$
h. $\begin{array}{r} 1 \\ +8 \\ \hline \end{array}$
i. $\begin{array}{r} 1 \\ +4 \\ \hline \end{array}$
j. $\begin{array}{r} 10 \\ +1 \\ \hline \end{array}$

Mastery Test 3

Name _____

a. $10 - 10$ d. $7 - 0$ g. $10 - 1$

b. $7 - 1$ e. $6 - 1$ h. $7 - 7$

c. $9 - 9$ f. $9 - 0$ i. $10 - 0$

Part 7

a. $70 + 10$
   $72 + 10$

b. $20 + 10$
   $25 + 10$

c. $50 + 10$
   $53 + 10$

d. $30 + 10$
   $36 + 10$

Part 8

a. $\begin{array}{r} 46 \\ -10 \\ \hline \end{array}$ c. $\begin{array}{r} 75 \\ +21 \\ \hline \end{array}$

b. $\begin{array}{r} 51 \\ +23 \\ \hline \end{array}$ d. $\begin{array}{r} 94 \\ -11 \\ \hline \end{array}$

Independent Seatwork

a. $40 + 6$ d. $60 + 8$ g. $30 + 1$

b. $10 + 9$ e. $20 + 5$ h. $80 + 2$

c. $90 + 3$ f. $70 + 4$ i. $50 + 7$

Copyright © The McGraw-Hill Companies, Inc. Permission is granted to reproduce for classroom use.

Remedies

Name _____

Copyright © The McGraw-Hill Companies, Inc.

Part A

a. 5 1, ⟶ __

b. 7 1, ⟶ __

c. 9 1, ⟶ __

_____   _____   _____

_____   _____   _____

Part B

a. _____   b. _____   c. _____   d. _____

Part C

a. _____   b. _____   c. _____

Part D

a. _____   b. _____

Part E

a. 7 1, ⟶ __

Part F

b. 9 1, ⟶ __

c. 6 1, ⟶ __

_____   _____   _____

_____   _____   _____

Part G

a. ____ + ___ = 13   c. ____ + ___ = 15

b. ____ + ___ = 71   d. ____ + ___ = 51

Part H

a. ____ + ___ = 17   c. ____ + ___ = 90

b. ____ + ___ = 70   d. ____ + ___ = 19

Remedies CONTINUED

Name _____

Part I

a. ____ + __ = 20   c. ____ + __ = 15

b. ____ + __ = 50   d. ____ + __ = 12

Part J

a. $12 - 0$ d. $9 - 1$ g. $9 - 0$

b. $7 - 1$ e. $6 - 6$ h. $8 - 1$

c. $10 - 10$ f. $9 - 9$ i. $8 - 8$

Part K

a. $\begin{array}{r} 19 \\ + 1 \\ \hline \end{array}$ b. $\begin{array}{r} 40 \\ - 1 \\ \hline \end{array}$ c. $\begin{array}{r} 50 \\ + 7 \\ \hline \end{array}$

Part L

a. $\begin{array}{r} 46 \\ - 10 \\ \hline \end{array}$ b. $\begin{array}{r} 75 \\ + 21 \\ \hline \end{array}$ c. $\begin{array}{r} 84 \\ - 14 \\ \hline \end{array}$

Part M

a. $\begin{array}{r} 51 \\ + 23 \\ \hline \end{array}$ b. $\begin{array}{r} 94 \\ - 11 \\ \hline \end{array}$ c. $\begin{array}{r} 26 \\ + 50 \\ \hline \end{array}$

Connecting Math Concepts

Copyright © The McGraw-Hill Companies, Inc.

Mastery Test 4

Copyright © The McGraw-Hill Companies, Inc. Permission is granted to reproduce for classroom use.

Name _____

Part 1

a. _____

b. _____

c. _____

d. _____

Part 2

a. $8 \longrightarrow 9$

b. $7 \quad 1 \longrightarrow \underline{}$

c. $\underline{} \quad 1 \longrightarrow 10$

d. $6 \longrightarrow 7$

e. $5 \quad 1 \longrightarrow \underline{}$

f. $\underline{} \quad 1 \longrightarrow 3$

Part 3

a.
$$\begin{array}{r} 357 \\ -\ 51 \\ \hline \end{array}$$

b.
$$\begin{array}{r} 51 \\ +628 \\ \hline \end{array}$$

c.
$$\begin{array}{r} 946 \\ -810 \\ \hline \end{array}$$

d.
$$\begin{array}{r} 261 \\ +\quad 5 \\ \hline \end{array}$$

Part 4

a. $300 + 40 + 8$

b. $500 + 60 + 1$

c. $100 + 70 + 6$

d. $600 + 30 + 9$

Part 5

a. $7 - 1$

b. $1 + 7$

c. $8 - 7$

d. $1 + 8$

e. $4 - 3$

f. $4 + 1$

g. $8 - 1$

h. $1 + 5$

i. $10 - 9$

Remedies

Copyright © The McGraw-Hill Companies, Inc.

Name _____

Part A

a. _____

b. _____

c. _____

Part B

a. _____

b. _____

c. _____

d. _____

Part C

a.

b.

c.

d.

Part D

a. $\dfrac{\quad}{\quad}\overset{1}{\longrightarrow}7$

b. $\dfrac{\quad}{\quad}\overset{1}{\longrightarrow}4$

c. $\dfrac{\quad}{\quad}\overset{1}{\longrightarrow}9$

_____ _____ _____

Part E

a. $\dfrac{7}{\quad}\longrightarrow 9$

b.  $\dfrac{6\quad 1}{\quad}\longrightarrow \underline{\ }$

c. $\dfrac{\quad}{\quad}\overset{2}{\longrightarrow}8$

d. $\dfrac{5}{\quad}\longrightarrow 7$

Part F

a.
$$\begin{array}{r} 946 \\ -810 \\ \hline \end{array}$$

b.
$$\begin{array}{r} 217 \\ +531 \\ \hline \end{array}$$

Part G

a.
$$\begin{array}{r} 357 \\ -\ 51 \\ \hline \end{array}$$

b.
$$\begin{array}{r} 51 \\ +628 \\ \hline \end{array}$$

Connecting Math Concepts

Remedies CONTINUED

Name _____

Copyright © The McGraw-Hill Companies, Inc.

Part H

a. $\begin{array}{r} 261 \\ +5 \\ \hline \end{array}$ b. $\begin{array}{r} 574 \\ -70 \\ \hline \end{array}$ c. $\begin{array}{r} 326 \\ -125 \\ \hline \end{array}$

Part I

a. $500 + 20 + 6 = $ _____

b. $100 + 80 + 4 = $ _____

c. $600 + 30 + 9 = $ _____

Part J

a. $500 + 60 + 1 = $ _____

b. _____ + _____ + ____ $= 348$

c. _____ + _____ + ____ $= 176$

Part K

a. $8 - 1$ d. $6 - 1$ g. $4 + 1$

b. $4 - 3$ e. $1 + 8$ h. $1 + 5$

c. $6 + 1$ f. $10 - 9$

Part L

a. $1 + 4$ d. $7 - 1$ g. $7 - 6$

b. $9 - 8$ e. $8 - 7$ h. $1 + 7$

c. $5 + 1$ f. $1 + 8$

Remedies

Mastery Test 5

Name _____

Part 1

a. _____

b. _____

c. _____

d. _____

e. _____

Part 2

a. 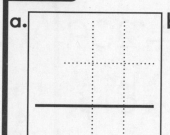 b. c.

Part 3

a. $8 \longrightarrow 9$

_____

b. $9 \quad 2 \longrightarrow \underline{\quad}$

_____

c. $\underline{\quad} \quad 2 \longrightarrow 9$

_____

d. $6 \longrightarrow 8$

_____

e. $5 \quad 1 \longrightarrow \underline{\quad}$

Part 4

a. $37 - 15$ c. $479 - 61$

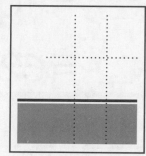   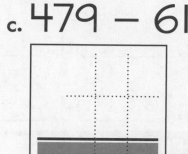

b. $17 + 62$ d. $50 + 318$

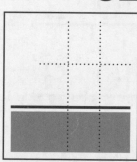   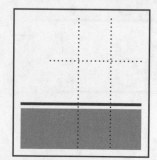

Part 5

a.

b.

c.

d.

Copyright © The McGraw-Hill Companies, Inc. Permission is granted to reproduce for classroom use.

Mastery Test 5

Name _____

Copyright © The McGraw-Hill Companies, Inc. Permission is granted to reproduce for classroom use.

Part 6

a. _____ + ___ + ___ = 209

b. _____ + ___ + ___ = 813

c. _____ + ___ + ___ = 500

d. _____ + ___ + ___ = 406

e. _____ + ___ + ___ = 620

Part 7

a. $\underrightarrow{10 \quad 7}$ ___

b. $\underrightarrow{10 \quad 1}$ ___

c. $\underrightarrow{10 \quad 8}$ ___

d. $\underrightarrow{10 \quad 3}$ ___

e. $\underrightarrow{10 \quad 5}$ ___

f. $\underrightarrow{10 \quad 2}$ ___

g. $\underrightarrow{10 \quad 9}$ ___

h. $\underrightarrow{10 \quad 6}$ ___

i. $\underrightarrow{10 \quad 4}$ ___

Part 9

a. $5 + 2$

b. $9 - 7$

c. $6 - 2$

d. $2 + 6$

e. $7 - 5$

f. $8 - 2$

g. $9 + 2$

h. $5 - 3$

i. $2 + 7$

Part 8

a. $1 + 7 + 2$

b. $9 + 1 + 6$

c. $1 + 4 + 2$

d. $1 + 9 + 5$

Connecting Math Concepts

Remedies

Name _____

Copyright © The McGraw-Hill Companies, Inc.

Part A

a. 207   b. 62   c. 17   d. 835

	H	T	O
a.			
b.			
c.			
d.			

Part B

a.		
b.		
c.		
d.		
e.		

Part C

a. 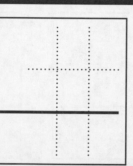   b.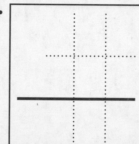

Part D

a. 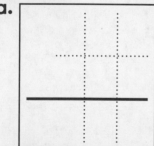   b. 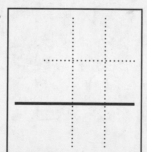   c.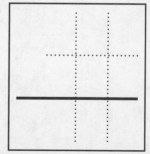

Part E

a. 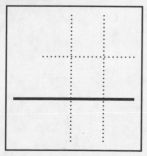   b. 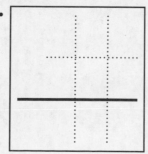   c.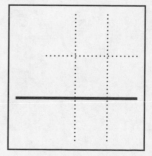

Remedies

Remedies CONTINUED

Name _____

Part F

a. $9 \longrightarrow 11$

b. $\underline{\quad} \xrightarrow{1} 9$

c. $9 \xrightarrow{2} \underline{\quad}$

d. $\underline{\quad} \xrightarrow{2} 9$

e. $5 \longrightarrow 6$

f. $\underline{\quad} \xrightarrow{2} 4$

g. $4 \xrightarrow{2} \underline{\quad}$

h. $8 \longrightarrow 10$

Part G

a. $17 + 62$

b. $36 - 15$

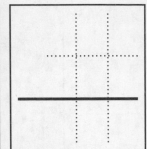
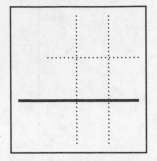

Part H

a. $50 + 318$

b. $479 - 61$

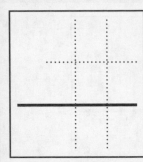
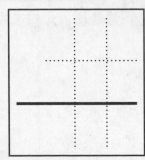

Part I

a. =

b. =

Part J

a. =

b. =

Copyright © The McGraw-Hill Companies, Inc.

Connecting Math Concepts

Remedies CONTINUED

Name _____

Part K

a. _____ + _____ + ___ = 782

b. _____ + _____ + ___ = 230

c. _____ + _____ + ___ = 193

d. _____ + _____ + ___ = 850

Part L

a. _____ + _____ + ___ = 506

b. _____ + _____ + ___ = 620

c. _____ + _____ + ___ = 209

d. _____ + _____ + ___ = 300

Part M

a. _____ + _____ + ___ = 102

b. _____ + _____ + ___ = 240

c. _____ + _____ + ___ = 711

Part N

a. _____ + _____ + ___ = 813

b. _____ + _____ + ___ = 500

c. _____ + _____ + ___ = 406

d. _____ + _____ + ___ = 170

Part O

a. $\underrightarrow{10 \quad 4}$ ___

b. $\underrightarrow{10 \quad 8}$ ___

c. $\underrightarrow{10 \quad 6}$ ___

d. $\underrightarrow{10 \quad 9}$ ___

e. $\underrightarrow{10 \quad 3}$ ___

f. $\underrightarrow{10 \quad 5}$ ___

Copyright © The McGraw-Hill Companies, Inc.

Remedies

Remedies CONTINUED

Name _____

Part P

a. $\underline{10 \quad 3}$, __

b. $\underline{10 \quad 7}$, __

c. $\underline{10 \quad 5}$, __

d. $\underline{10 \quad 2}$, __

e. $\underline{10 \quad 6}$, __

f. $\underline{10 \quad 1}$, __

Part Q

a. $\underline{3 + 2} + 1$

b. $\underline{7 + 1} + 2$

Part R

a. $1 + 9 + 5$

b. $1 + 5 + 1$

Part S

a. $9 - 7$

b. $5 + 2$

c. $6 - 2$

d. $7 - 5$

e. $2 + 7$

f. $8 - 2$

Part T

a. $9 - 8$

b. $6 - 1$

c. $5 - 3$

d. $1 + 4$

e. $8 - 7$

f. $3 - 2$

g. $1 + 10$

h. $7 - 1$

i. $2 + 7$

j. $6 - 4$

Copyright © The McGraw-Hill Companies, Inc.

Connecting Math Concepts

Mastery Test 6

Name _____

Part 1

a.

b.

c.

Part 2

a. $58 + $ ___ $=$

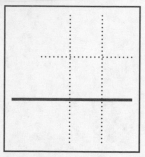

d. $73 + $ ___ $=$

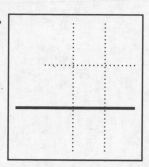
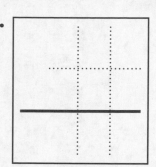

b. $17 + $ ___ $=$

e. $37 + $ ___ $=$

c. $28 + $ ___ $=$

Part 3

a. $6 - 4$

b. $9 + 2$

c. $13 - 3$

d. $11 - 2$

e. $8 + 10$

f. $8 - 2$

g. $10 - 8$

h. $2 + 6$

i. $17 - 7$

j. $4 - 2$

Copyright © The McGraw-Hill Companies, Inc. Permission is granted to reproduce for classroom use.

Mastery Test 6

Name _____

Part 4

Part 5

a. $8-3$

b. $10-6$

c. $3+5$

d. $9-3$

e. $6+4$

f. $9-6$

g. $8-5$

h. $3+6$

i. $10-4$

j. $5+3$

Part 6

a. $\dfrac{\;\;\;}{+\;\;\;}$ 382

b. $\dfrac{\;\;\;}{+\;\;\;}$ 107

c. $\dfrac{\;\;\;}{+\;\;\;}$ 619

d. $\dfrac{\;\;\;}{+\;\;\;}$ 760

Part 7

a. $\begin{array}{r} 265 \\ -210 \\ \hline \end{array}$

b. $\begin{array}{r} 265 \\ +210 \\ \hline \end{array}$

c. $\begin{array}{r} 468 \\ -426 \\ \hline \end{array}$

d. $\begin{array}{r} 364 \\ -\;\;54 \\ \hline \end{array}$

Connecting Math Concepts

Copyright © The McGraw-Hill Companies, Inc. Permission is granted to reproduce for classroom use.

Remedies

Name _____

Part A

a.    b.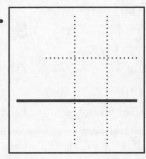

Part B

a. 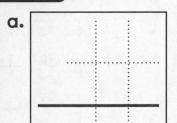   b.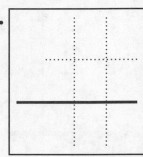

Part C

a. ___ + ___ =

b. ___ + ___ =

c. ___ + ___ =

Part D

a. 18   ___ =

b. 37   ___ =

Part E

a. 34   ___ =

b. 17   ___ =

c. 58   ___ =

d. 66   ___ =

Copyright © The McGraw-Hill Companies, Inc.

Part F

a. $2+5$ d. $6+2$ g. $3+2$

b. $2+7$ e. $8+2$ h. $5+3$

c. $3+5$ f. $3+10$ i. $2+10$

Part G

a.  =   b. =

c. =

Part H

a. =

b. =

Part I

a.  =   b.  =

Part J

d. $10+8$

a. $6+4$ e. $2+3$ h. $4+6$

b. $7+2$ f. $5+10$ i. $5+2$

c. $3+6$ g. $2+9$ j. $3+5$

Copyright © The McGraw-Hill Companies, Inc.

Remedies

Remedies CONTINUED

Name _____

Part K

a.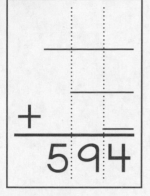
+ 594

b.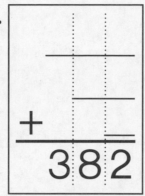
+ 382

c.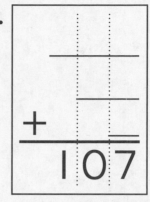
+ 107

Part L

a.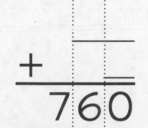
+ 760

b.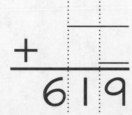
+ 619

c.
+ 181

d.
+ 503

Part M

a.
```
  58
- 57
```

b.
```
  364
- 343
```

c.
```
  496
- 416
```

Part N

a.
```
  725
- 705
```

b.
```
   53
+ 821
```

c.
```
  96
- 91
```

d.
```
  364
-  54
```

Copyright © The McGraw-Hill Companies, Inc.

Cumulative Test 1

Name _____

Part 1

a. _____   b. _____   c. _____   d. _____   e. _____

f. _____   g. _____   h. _____   i. _____   j. _____

Part 2

a. $2 + 1$

b. $12 + 1$

c. $6 + 1$

d. $16 + 1$

e. $3 + 1$

f. $13 + 1$

g. $1 + 1$

h. $11 + 1$

i. $4 + 1$

j. $14 + 1$

k. $8 + 1$

l. $8 + 2$

m. $9 + 1$

n. $9 + 2$

o. $5 + 1$

p. $5 + 2$

q. $7 + 1$

r. $7 + 2$

s. $10 + 1$

t. $10 + 2$

u. $4 + 0$

v. $1 + 0$

w. $10 + 0$

x. $38 + 0$

y. $0 + 0$

Part 3

$$\underline{90} + \underline{6} = 96$$

a. _____ + ___ = 37

b. _____ + ___ = 81

c. _____ + ___ = 90

d. _____ + ___ = 19

e. _____ + ___ = 15

f. _____ + ___ = 50

Copyright © The McGraw-Hill Companies, Inc.

Cumulative Test 1

Name _____

Part 4

a. _____

b. _____

c. _____

d. _____

e. _____

f. _____

g. _____

h. _____

Part 5

a. _____

b. _____

c.

d.

e.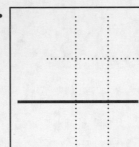

f.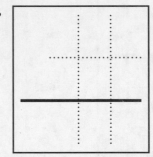

Part 6

a. $300 + 40 + 8 =$

b. $500 + 60 + 1 =$

c. $100 + 70 + 6 =$

d. ____+____+___=209

e. ____+____+___=813

f. ____+____+___=620

g.
$$+ \underline{} \over 619$$

h.
$$+ \underline{} \over 760$$

Part 7

a. $2 + 7 = 9$ b. $30 + 8 = 38$ c. $15 + 1 = 16$

Copyright © The McGraw-Hill Companies, Inc.

Cumulative Test 1

Name _____

Part 8

a. $5 \quad 1 \longrightarrow 6$

_____
_____
_____

b. $7 \quad 2 \longrightarrow 9$

_____
_____
_____

Part 9

a.
$$\begin{array}{r} 46 \\ -\ 10 \\ \hline \end{array}$$

b.
$$\begin{array}{r} 51 \\ +\ 23 \\ \hline \end{array}$$

c.
$$\begin{array}{r} 75 \\ +\ 21 \\ \hline \end{array}$$

d.
$$\begin{array}{r} 357 \\ -\ 51 \\ \hline \end{array}$$

e.
$$\begin{array}{r} 51 \\ +628 \\ \hline \end{array}$$

f.
$$\begin{array}{r} 946 \\ -810 \\ \hline \end{array}$$

g.
$$\begin{array}{r} 265 \\ -210 \\ \hline \end{array}$$

h.
$$\begin{array}{r} 265 \\ +210 \\ \hline \end{array}$$

i.
$$\begin{array}{r} 468 \\ -426 \\ \hline \end{array}$$

j.
$$\begin{array}{r} 364 \\ -\ 54 \\ \hline \end{array}$$

Part 10

a. $8 \longrightarrow 9$

_____

b. $9 \quad 2 \longrightarrow \underline{\ \ }$

_____

c. $\underline{\ \ } \quad 2 \longrightarrow 9$

_____

d. $6 \longrightarrow 8$

e. $5 \quad 1 \longrightarrow \underline{\ \ }$

_____

Cumulative Test 1

Name _____

Part 11

a. $58 + \underline{\hphantom{00}} =$

c. $28 + \underline{\hphantom{00}} =$

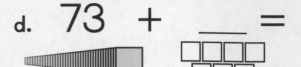

b. $17 + \underline{\hphantom{00}} =$

d. $73 + \underline{\hphantom{00}} =$

Part 12

a. $7 - 1$

b. $1 + 7$

c. $8 - 7$

d. $1 + 8$

e. $4 - 3$

f. $4 + 1$

g. $8 - 1$

h. $1 + 5$

i. $10 - 9$

j. $5 + 2$

k. $9 - 7$

l. $6 - 2$

m. $2 + 6$

n. $7 - 5$

o. $8 - 2$

p. $9 + 2$

q. $5 - 3$

r. $2 + 7$

s. $13 - 3$

t. $11 - 1$

u. $8 + 10$

v. $10 - 8$

w. $2 + 10$

x. $17 - 7$

y. $4 - 2$

Part 13

a. $9 - 1$

b. $6 - 1$

c. $3 - 1$

d. $8 - 8$

e. $2 - 1$

f. $5 - 1$

g. $10 - 10$

h. $7 - 1$

i. $10 - 9$

j. $7 - 0$

k. $7 - 6$

l. $9 - 8$

m. $10 - 1$

n. $7 - 7$

o. $10 - 0$

Copyright © The McGraw-Hill Companies, Inc.

Connecting Math Concepts

Cumulative Test 1

Name _____

Part 14

a.

b.

c.

d.

Part 15

a. $70 + 10$

b. $72 + 10$

c. $20 + 10$

d. $25 + 10$

e. $50 + 10$

f. $53 + 10$

g. $30 + 10$

h. $36 + 10$

Part 16

a. $1 + 7 + 2$

b. $9 + 1 + 6$

c. $1 + 4 + 2$

d. $1 + 9 + 5$

Part 17

a. $8 - 3$

b. $10 - 6$

c. $3 + 5$

d. $9 - 3$

e. $6 + 4$

f. $9 - 6$

g. $8 - 5$

h. $3 + 6$

i. $10 - 4$

j. $5 + 3$

Copyright © The McGraw-Hill Companies, Inc.

Mastery Test 7

Part 1

a. $\underline{10} \xrightarrow{\ 6\ } \underline{\ \ }$

b. $\underline{\ \ } \xrightarrow{\ 5\ } 11$

c. $\underline{6} \xrightarrow{\ 6\ } \underline{\ \ }$

d. $\underline{\ \ } \xrightarrow{\ 2\ } 6$

e. $\underline{6} \xrightarrow{\ 5\ } \underline{\ \ }$

f. $\underline{6} \xrightarrow{\ \ } 7$

g. $\underline{6} \xrightarrow{\ 2\ } \underline{\ \ }$

h. $\underline{\ \ } \xrightarrow{\ 6\ } 12$

i. $\underline{6} \xrightarrow{\ \ } 8$

j. $\underline{6} \xrightarrow{\ \ } 11$

Part 2

a. $\underline{\ \ } + \underline{\ \ } =$

b. $\underline{\ \ } + \underline{\ \ } =$

c. $47 + \underline{\ \ } =$

d. $15 + \underline{\ \ } =$

e. $54 + \underline{\ \ } =$

Part 3

a. $20, 10, 5, 5$

b. $50, 10, 5, 1, 1$

c. $10, 10, 5, 5, 5$

d. $50, 10, 10, 5, 5$

e. $20, 10, 10, 5, 1, 1$

Copyright © The McGraw-Hill Companies, Inc. Permission is granted to reproduce for classroom use.

Mastery Test 7

Name _____

Part 4

a.
```
  34
  21
+ 23
```

b.
```
  50
  32
+ 12
```

c.
```
  32
  26
+ 21
```

d.
```
  18
  20
+ 40
```

e.
```
  274
   12
+ 102
```

Part 5

a.

b.

c.

d.

e.

Part 6

a.
```
  857
-  37
```

c.
```
  459
- 402
```

e.
```
   98
-  96
```

b.
```
  386
- 381
```

d.
```
  738
- 532
```

Copyright © The McGraw-Hill Companies, Inc. Permission is granted to reproduce for classroom use.

Connecting Math Concepts

Remedies

Name _____

Copyright © The McGraw-Hill Companies, Inc.

Part A

$$3 + 6 =$$

a.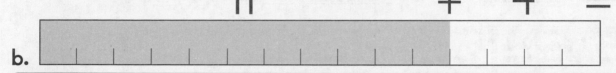

$$11 + 4 =$$

b.

Part B

$$15 + \underline{\quad} =$$

a.

$$47 + \underline{\quad} =$$

b.

Part C

$$21 + \underline{\quad} =$$

a.

$$24 + \underline{\quad} =$$

b.

Part D

a.

b.

c.

Remedies

Remedies CONTINUED

Part E

a. [10] [10] [5] [5] [1]

b. [20] [5] [5] [10] [10]

c. [50] [10] [10] [5] [5] [5]

Part F

a.
```
  34
  21
+ 23
```

b.
```
  12
  11
+ 32
```

c.
```
  50
  34
+ 12
```

Part G

a.
```
  25
  30
+ 32
```

b.
```
   18
  201
+  40
```

c.
```
  356
+ 620
```

d.
```
   61
+ 320
```

Copyright © The McGraw-Hill Companies, Inc.

Remedies

Remedies CONTINUED

Name _____

Part H

a. (2 quarters)
_____

b. (4 quarters)
_____

c. (1 quarter, 3 nickels)
_____

Part I

a. (3 quarters, 3 nickels)
_____

b. (2 quarters, 3 dimes)
_____

c. (1 quarter, 3 dimes)
_____

Part J

a. (3 quarters, 3 pennies)
_____

b. (3 quarters, 3 nickels)
_____

c. (2 quarters, 2 dimes, 2 nickels)
_____

Copyright © The McGraw-Hill Companies, Inc.

Mastery Test 8

Name _____

Copyright © The McGraw-Hill Companies, Inc. Permission is granted to reproduce for classroom use.

Part 1

a. _____   b. _____   c. _____   d. _____   e. _____

Part 2

a. _____ + ___ = 19

c. _____ + ___ = 21

b. _____ + ___ = 37

d. _____ + ___ = 62

Part 3

a. $3 + 6$

b. $2 + 9$

c. $6 + 6$

d. $7 + 2$

e. $3 + 5$

f. $6 + 4$

g. $2 + 8$

h. $6 + 5$

i. $2 + 10$

j. $6 + 3$

Part 4

a. = ☐ one dollar

b. = ☐ one dollar

c. = ☐ one dollar

d. = ☐ one dollar

e. = ☐ one dollar

Mastery Test 8

Part 5

a. $11 - 5$

b. $11 - 9$

c. $9 - 3$

d. $9 - 7$

e. $8 - 5$

f. $12 - 6$

g. $12 - 10$

h. $10 - 4$

i. $10 - 2$

j. $11 - 6$

Part 6

a.
$$\begin{array}{r} 72 \\ + 18 \\ \hline \end{array}$$

b.
$$\begin{array}{r} 16 \\ + 46 \\ \hline \end{array}$$

c.
$$\begin{array}{r} 76 \\ + 14 \\ \hline \end{array}$$

d.
$$\begin{array}{r} 8 \\ 52 \\ + 37 \\ \hline \end{array}$$

Part 7

a.
$$\begin{array}{r} \$4.25 \\ -3.15 \\ \hline \end{array}$$

b.
$$\begin{array}{r} \$5.25 \\ +2.41 \\ \hline \end{array}$$

c.
$$\begin{array}{r} \$\ \ 3.26 \\ +11.52 \\ \hline \end{array}$$

d.
$$\begin{array}{r} \$12.41 \\ -10.31 \\ \hline \end{array}$$

Copyright © The McGraw-Hill Companies, Inc. Permission is granted to reproduce for classroom use.

Remedies

Name _____

Part A

a. $ _____ b. $ _____ c. $ _____

Part B

a. _____ b. _____ c. _____

Part C

a. _____ b. _____ c. _____ d. _____

Part D

a. ____ + 5 = 17

b. ____ + 8 = 16

c. ____ + 5 = 14

Part E

a. ____ + ___ = 37

b. ____ + ___ = 21

Part F

a. ____ + ___ = 62

b. ____ + ___ = 19

Part G

a. 6 + 4

b. 2 + 9

c. 3 + 6

d. 5 + 2

e. 6 + 6

f. 6 + 5

g. 8 + 2

h. 3 + 5

i. 2 + 6

j. 9 + 1

Copyright © The McGraw-Hill Companies, Inc.

Remedies

Remedies CONTINUED

Name _____

Part H

a. $7 + 1$

b. $4 + 6$

c. $5 + 3$

d. $2 + 7$

e. $6 + 6$

f. $4 + 2$

g. $3 + 6$

h. $6 + 10$

i. $5 + 6$

j. $6 + 2$

Part I

a. = ☐ one dollar

b. = ☐ one dollar

c. = ☐ one dollar

d. = ☐ one dollar

e. = ☐ one dollar

Part J

a. $10 - 6$

b. $11 - 9$

c. $12 - 2$

d. $12 - 6$

e. $8 - 5$

f. $11 - 5$

g. $8 - 2$

h. $10 - 8$

i. $9 - 3$

j. $6 - 4$

Connecting Math Concepts

Copyright © The McGraw-Hill Companies, Inc.

Remedies CONTINUED

Name _____

Part K

a. $3 - 2$ e. $6 - 2$ i. $8 - 5$

b. $12 - 6$ f. $8 - 6$ j. $11 - 2$

c. $14 - 10$ g. $9 - 7$ k. $11 - 5$

d. $4 - 1$ h. $10 - 6$ l. $9 - 6$

Part L

a.
$$\begin{array}{r} 34 \\ 12 \\ +\ 36 \\ \hline \end{array}$$

b.
$$\begin{array}{r} 8 \\ 52 \\ +\ 37 \\ \hline \end{array}$$

c.
$$\begin{array}{r} 76 \\ +\ 14 \\ \hline \end{array}$$

Part M

a.
$$\begin{array}{r} \$5.25 \\ +\ 2.41 \\ \hline \end{array}$$

b.
$$\begin{array}{r} \$4.25 \\ -\ 3.15 \\ \hline \end{array}$$

Part N

a.
$$\begin{array}{r} \$7.30 \\ +\ 2.27 \\ \hline \end{array}$$

b.
$$\begin{array}{r} \$3.89 \\ -\ 2.37 \\ \hline \end{array}$$

c.
$$\begin{array}{r} \$12.41 \\ -\ 10.31 \\ \hline \end{array}$$

d.
$$\begin{array}{r} \$\ \ 3.26 \\ +\ 11.52 \\ \hline \end{array}$$

Copyright © The McGraw-Hill Companies, Inc.

Mastery Test 9

Name _____

Copyright © The McGraw-Hill Companies, Inc. Permission is granted to reproduce for classroom use.

Part 1

a. 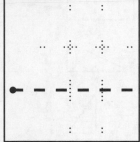 b. 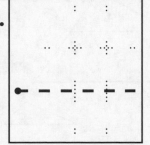 c.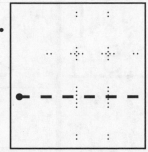

Part 2

a. _____ •   b. _____ •   c. _____ •   d. _____ •   e. _____ •

Part 3

a.
4 feet
_____
10 feet

c.
27 feet
_____
16 feet
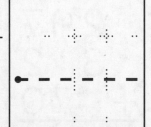

b.
18 feet
_____
8 feet
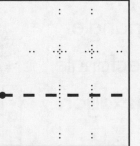

d.
13 feet
_____
45 feet
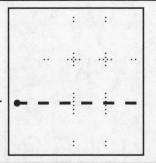

Part 4

a. 7   5   e. 3   2

b. 8   9   f. 19   20

c. 6   3   g. 57   48

d. 4   4   h. 16   16

Mastery Test 9

Name _____

Part 5

a.
$$\begin{array}{r} 1\,1 \\ +\ 5\,9 \\ \hline \end{array}$$

b.
$$\begin{array}{r} 3\,7 \\ +\ 4\,2 \\ \hline \end{array}$$

c.
$$\begin{array}{r} 2\,9\,7 \\ 3\,1\,2 \\ +\ 7\,0 \\ \hline \end{array}$$

d.
$$\begin{array}{r} 4\,1\,2 \\ 8 \\ +1\,5\,3 \\ \hline \end{array}$$

e.
$$\begin{array}{r} 8\,1 \\ 1\,2\,4 \\ +5\,6\,2 \\ \hline \end{array}$$

Part 6

1.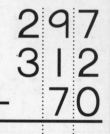

Triangle
Rectangle
Hexagon

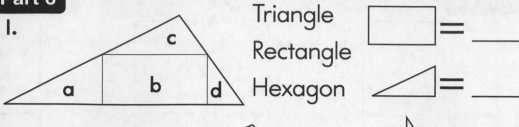

□ = ____
△ = ____
△ = ____   ◿ = ____

2.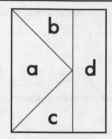

Triangle
Rectangle
Hexagon

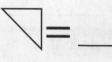

▯ = ____   ◸ = ____
▷ = ____   ◹ = ____

3.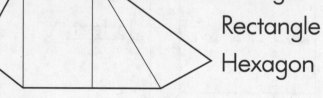

Triangle
Rectangle
Hexagon

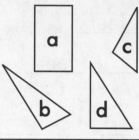

4.

Triangle
Rectangle
Hexagon

Part 7

a. $4 + 3$
b. $4 - 3$
c. $4 + 6$
d. $8 - 4$

e. $9 - 3$
f. $4 + 4$
g. $7 - 4$
h. $10 - 4$
i. $7 - 3$

Connecting Math Concepts

Copyright © The McGraw-Hill Companies, Inc. Permission is granted to reproduce for classroom use.

Remedies

Name _____

Part A

a. 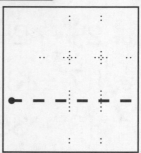 b.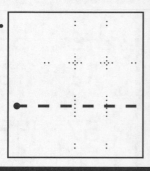

Part B

a. 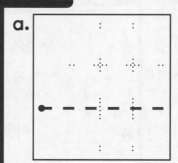 b.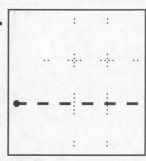

Part C

a. $18.____   b. $18.____   c. $18.____

Part D

a. $__.__   b. $__.__   c. $__.__

Part E

a. $\dfrac{8}{6}$

b. $\dfrac{4}{10}$

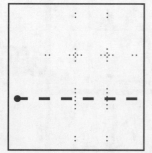

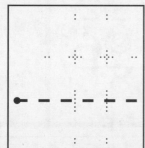

Part F

a. $\dfrac{\text{8 feet}}{\text{18 feet}}$

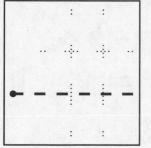

b. $\dfrac{\text{45 miles}}{\text{13 miles}}$

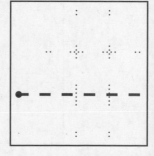

Copyright © The McGraw-Hill Companies, Inc.

Remedies CONTINUED

Name _____

Part G

a. <u>   27 feet   </u>

<u> 16 feet </u>

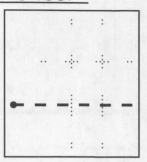

b. <u>24 meters</u>

<u> 64 meters </u>

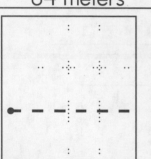

Part H

a. 19 ✕ 20

b. 33 ✕ 22

c. 53 ✕ 48

Part I

a. 46   53   d. 24   24

b. 16   16   e. 117   130

c. 57   48   f. 43   35

Part J

a. $\begin{array}{r} 11 \\ +59 \\ \hline \end{array}$
b. $\begin{array}{r} 37 \\ +42 \\ \hline \end{array}$
c. $\begin{array}{r} 38 \\ +42 \\ \hline \end{array}$

Part K

a.

$\begin{array}{r} 122 \\ +486 \\ \hline \end{array}$

Part L

a. $\begin{array}{r} 297 \\ 12 \\ +370 \\ \hline \end{array}$
b. $\begin{array}{r} 8 \\ 412 \\ +153 \\ \hline \end{array}$

Part M

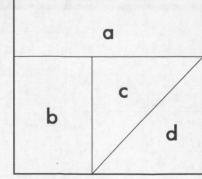

Copyright © The McGraw-Hill Companies, Inc.

Remedies CONTINUED

Name _____

Part N

1.

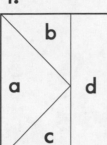

Hexagon
Rectangle
Triangle

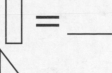

 = _____

= _____

= _____

= _____

2.

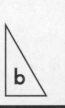

Hexagon
Rectangle
Triangle

 = _____

= _____

= _____

= _____

Part O

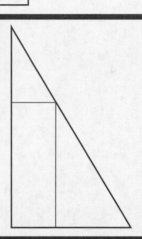

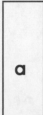

Part P

a. $8 - 4$

b. $3 - 2$

c. $9 - 6$

d. $6 - 2$

e. $8 - 6$

f. $11 - 6$

g. $7 - 4$

h. $12 - 6$

i. $11 - 9$

j. $7 - 3$

k. $8 - 3$

l. $7 - 5$

Part Q

a. $5 + 2$

b. $3 + 6$

c. $4 + 4$

d. $5 + 3$

e. $2 + 7$

f. $3 + 4$

g. $8 + 10$

h. $4 + 6$

i. $4 + 2$

j. $6 + 6$

k. $9 + 1$

l. $5 + 6$

Copyright © The McGraw-Hill Companies, Inc.

Connecting Math Concepts

Mastery Test 10

Name _____

a. _____ :   b. _____ :   c. _____ :   d. _____ :

e. _____ :   f. _____ :   g. _____ :   h. _____ :

Part 2

a. 24 ⟶ 35

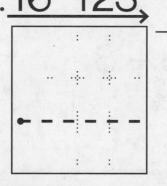

b. 16 423 ⟶ ___

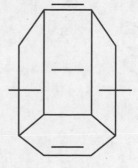

c. ___ ⟶ 72 ⟶ 93

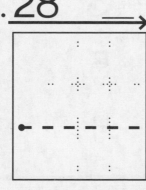

d. 28 ⟶ 79

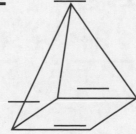

e. ___ ⟶ 115, 248

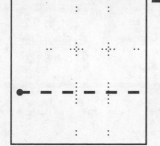

f. 104 90 ⟶ ___

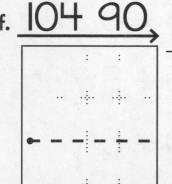

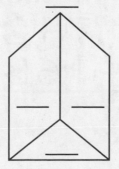

Part 3

1   2   3

Copyright © The McGraw-Hill Companies, Inc. Permission is granted to reproduce for classroom use.

Mastery Test 10 Name _____

Part 4

a. $61 +

b. $33 +

c.

d.

Part 5

a. $6 - 3$

b. $6 + 3$

c. $10 + 5$

d. $10 - 5$

e. $5 - 5$

f. $5 + 5$

g. $10 - 4$

h. $8 - 4$

i. $3 + 3$

j. $8 - 3$

Part 6

a.
$$742 + 168$$

b. 31
$$349 + 4$$

c. 261
$$6 + 62$$

d.
$$358 + 442$$

Connecting Math Concepts

Copyright © The McGraw-Hill Companies, Inc. Permission is granted to reproduce for classroom use.

Remedies

Copyright © The McGraw-Hill Companies, Inc.

Name _____

Part A

a. _____ :

b. _____ :

c. _____ :

d. _____ :

Part B

a. _____ :

b. _____ :

c. _____ :

d. _____ :

e. _____ :

f. _____ :

Part C

a. _____ :

b. _____ :

c. _____ :

d. _____ :

e. _____ :

f. _____ :

Part D

a. 51   36 ⟶ __

b. 17   ⟶ 29

Part E

a. 32   12 ⟶ __

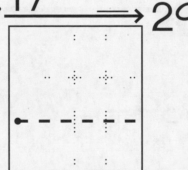

b. 24   ⟶ 35

c.   72 ⟶ 93

d. 50   27 ⟶ __

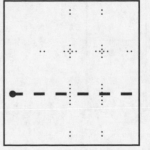

Remedies CONTINUED

Name _____

Part F

a. 28 ⟶ 79

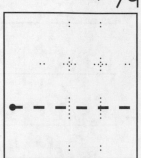

c. _____ 115 ⟶ 248

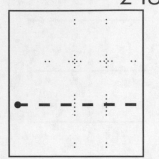

b. 16   423 ⟶ _____

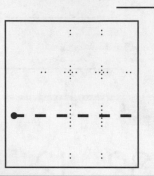

d. 104   90 ⟶ _____

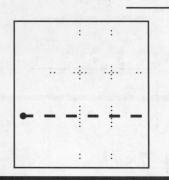

Part G

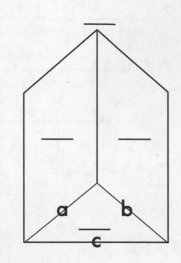

Part H

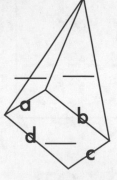

1

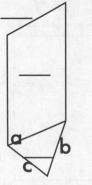

2

Part I

a. $39 + 🪙🪙🪙

b. $72 + 🪙🪙🪙🪙

c. $15 + 🪙🪙🪙🪙

d. $28 + 🪙🪙🪙🪙🪙

Copyright © The McGraw-Hill Companies, Inc.

Connecting Math Concepts

Remedies CONTINUED

Name _____

Part J

a.

b.

c.

d.

Part K

a. $6 - 3$
b. $5 + 5$
c. $8 - 5$

d. $3 + 4$
e. $10 - 6$
f. $5 - 3$
g. $3 + 3$

h. $10 - 5$
i. $5 + 10$
j. $8 - 4$

Part L

a. $11 - 5$
b. $6 + 3$
c. $5 + 5$
d. $12 - 6$
e. $10 - 5$

f. $4 + 3$
g. $6 - 3$
h. $6 + 2$
i. $8 - 4$
j. $4 + 6$

Part M

a.
$$\begin{array}{r} 186 \\ +716 \\ \hline \end{array}$$

b.
$$\begin{array}{r} 296 \\ +475 \\ \hline \end{array}$$

Part N

a.
$$\begin{array}{r} 457 \\ +162 \\ \hline \end{array}$$

b.
$$\begin{array}{r} 341 \\ 9 \\ +\ 34 \\ \hline \end{array}$$

c.
$$\begin{array}{r} 742 \\ +168 \\ \hline \end{array}$$

Copyright © The McGraw-Hill Companies, Inc.

Mastery Test 11

Name _____

Part 1

1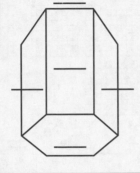

2

3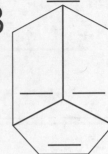

Part 2

a.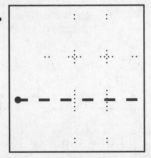

b.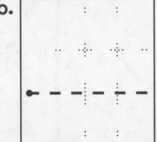

c.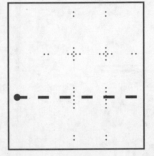

Part 3

a.

b.

c.

d.

e.

f.

g.

h.

Copyright © The McGraw-Hill Companies, Inc. Permission is granted to reproduce for classroom use.

Mastery Test 11

Name _____

Part 4

a.
$$\begin{array}{r} 220 \\ -142 \\ \hline 78 \end{array}$$

═══════⟶ ___

b.
$$\begin{array}{r} 181 \\ -142 \\ \hline 39 \end{array}$$

═══════⟶ ___

c.
$$\begin{array}{r} 220 \\ +181 \\ \hline 401 \end{array}$$

═══════⟶ ___

d.
$$\begin{array}{r} 142 \\ +181 \\ \hline 323 \end{array}$$

═══════⟶ ___

e.
$$\begin{array}{r} 142 \\ -\ 39 \\ \hline 103 \end{array}$$

═══════⟶ ___

Part 5

a. $1 + 3$

b. $2 + 3$

c. $3 + 3$

d. $4 + 3$

e. $5 + 3$

f. $6 + 3$

g. $7 + 3$

h. $8 + 3$

i. $9 + 3$

j. $10 + 3$

Part 7

a. $16 - 8$

b. $16 - 10$

c. $7 + 2$

d. $7 + 7$

e. $8 - 4$

f. $8 + 8$

g. $10 - 5$

Part 6

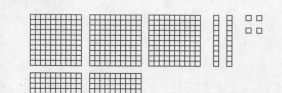

a.

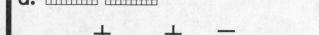
___ + ___ + ___ = ___

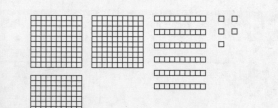

b.

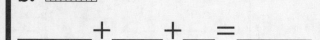
___ + ___ + ___ = ___

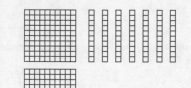

c.

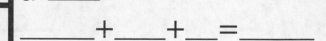
___ + ___ + ___ = ___

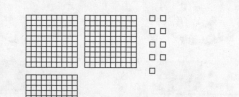

d.

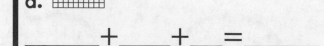
___ + ___ + ___ = ___

h. $6 - 3$

i. $6 + 6$

j. $14 - 7$

Connecting Math Concepts

Copyright © The McGraw-Hill Companies, Inc. Permission is granted to reproduce for classroom use.

Remedies

Name _____

Part A

1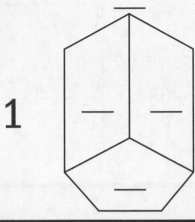

2

Part B

1 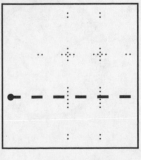   2 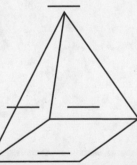   1 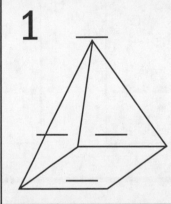   2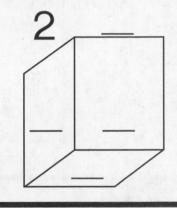

Part C

1 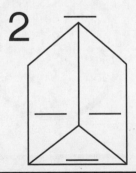   2 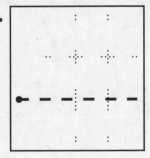   3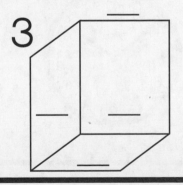

Part D

a.

b.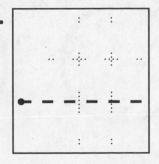

c.

Copyright © The McGraw-Hill Companies, Inc.

Remedies CONTINUED

Name _____

Part E

a. 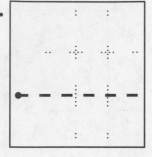   b. 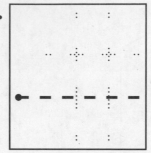   c.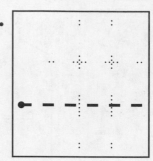

Part F

a. 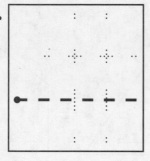   b. 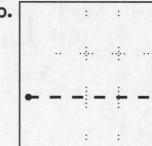   c.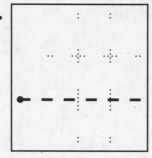

Part G

a.    b.    c.

Part H

a.    b.    c.    d.

Copyright © The McGraw-Hill Companies, Inc.

Connecting Math Concepts

Remedies CONTINUED

Name _____

Part I

a.

b.

c.

d.

_____ : ▮   _____ : ▮   _____ : ▮   _____ : ▮

Part J

a.
$$\begin{array}{r} 214 \\ -\ 96 \\ \hline 118 \end{array}$$

b.
$$\begin{array}{r} 137 \\ +385 \\ \hline 522 \end{array}$$

c.
$$\begin{array}{r} 385 \\ -214 \\ \hline 171 \end{array}$$

d.
$$\begin{array}{r} 171 \\ +\ 96 \\ \hline 267 \end{array}$$

Part K

a.
$$\begin{array}{r} 142 \\ +181 \\ \hline 323 \end{array}$$

b.
$$\begin{array}{r} 181 \\ -142 \\ \hline 39 \end{array}$$

c.
$$\begin{array}{r} 220 \\ -142 \\ \hline 78 \end{array}$$

d.
$$\begin{array}{r} 220 \\ +181 \\ \hline 401 \end{array}$$

Copyright © The McGraw-Hill Companies, Inc.

Remedies

Remedies CONTINUED

Name _____

Remedies

Part L

a.

____ + __ + __ = ____

b.

____ + __ + __ = ____

Part M

a.

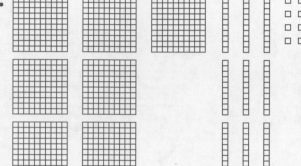

____ + __ + __ = ____

b.

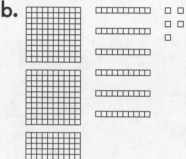

____ + __ + __ = ____

Part N

a.

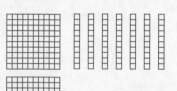

____ + __ + __ = ____

b.

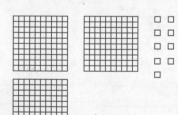

____ + __ + __ = ____

Connecting Math Concepts

Copyright © The McGraw-Hill Companies, Inc.

Remedies CONTINUED

Name _____

a. $10 - 4$ e. $8 + 8$ i. $6 - 3$

b. $4 + 4$ f. $10 - 5$ j. $6 + 3$

c. $12 - 6$ g. $7 + 7$ k. $10 - 6$

d. $8 - 4$ h. $14 - 7$

Part P

c. $6 + 3$ f. $10 - 5$

a. $6 + 4$ d. $11 - 6$ g. $7 + 7$

b. $6 - 2$ e. $3 + 4$ h. $8 - 4$

Copyright © The McGraw-Hill Companies, Inc.

Remedies

Mastery Test 12

Name _____

Copyright © The McGraw-Hill Companies, Inc. Permission is granted to reproduce for classroom use.

Part 1

a. _____ b. _____ c. _____ d. _____ e. _____

Part 2

a. ___ $- 18 = 63$

$\Longrightarrow$ ___

b. ___ $+ 27 = 68$

$\Longrightarrow$ ___

c. $68 - 23 =$ ___

$\Longrightarrow$ ___

d. $79 -$ ___ $= 18$

$\Longrightarrow$ ___

e. $54 +$ ___ $= 88$

$\Longrightarrow$ ___

f. ___ $- 23 = 27$

$\Longrightarrow$ ___

Part 3

a. _____

b. _____

Part 4

a. $184 + 108 =$ _____

b. _____ $- 184 = 76$

c. $184 +$ _____ $= 292$

d. $184 -$ _____ $= 76$

e. _____ $- 76 = 108$

f. _____ $+ 184 = 260$

Part 5

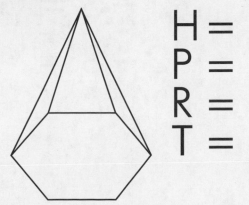

H =
P =
R =
T =

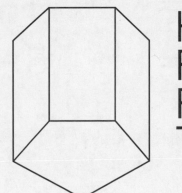

H =
P =
R =
T =

Mastery Test 12

Part 6

a. _____ − 23 = 65   b. _____ + 54 = 75   c. 89 − _____ = 63

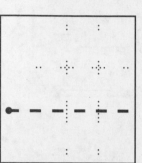 _____   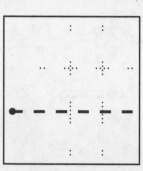 _____    _____

Part 7

a. 3 + 1 =

b. 3 + 2 =

c. 3 + 3 =

d. 3 + 4 =

e. 3 + 5 =

f. 3 + 6 =

g. 3 + 7 =

h. 3 + 8 =

i. 3 + 9 =

j. 3 + 10 =

k. 3 − 3 =

l. 4 − 3 =

m. 5 − 3 =

n. 6 − 3 =

o. 7 − 3 =

p. 8 − 3 =

q. 9 − 3 =

r. 10 − 3 =

s. 11 − 3 =

t. 12 − 3 =

Copyright © The McGraw-Hill Companies, Inc. Permission is granted to reproduce for classroom use.

Mastery Test 12

Name _____

Copyright © The McGraw-Hill Companies, Inc. Permission is granted to reproduce for classroom use.

Part 8

a.

_____:_____

c.

_____:_____

e.

_____:_____

g.

_____:_____

b.

_____:_____

d.

_____:_____

f.

_____:_____

h.

_____:_____

Part 9

a.   _____

b. _____

c. _____

d. 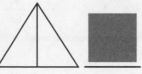 _____

Part 10

a. $10 - 7$

b. $8 + 8$

c. $8 + 3$

d. $8 - 3$

e. $11 - 3$

f. $7 - 3$

g. $7 + 3$

h. $7 + 7$

i. $11 - 8$

j. $10 - 3$

Remedies

Name _____

Part A

a.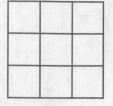

b.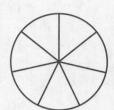

c.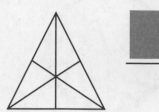

Part B

a.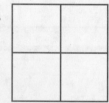

c.

e.

b.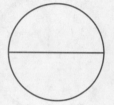

d.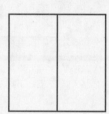

f.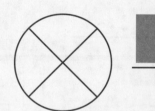

Part C

a. $79 - \underline{\quad} = 18$ c. $\underline{\quad} - 63 = 18$ e. $\underline{\quad} - 63 = 32$

b. $\underline{\quad} + 31 = 81$ d. $58 - \underline{\quad} = 17$ f. $18 + \underline{\quad} = 89$

Copyright © The McGraw-Hill Companies, Inc.

Remedies

Name _____

Part D

a. _____ + 27 = 68   c. 54 + _____ = 88   e. 88 – _____ = 23

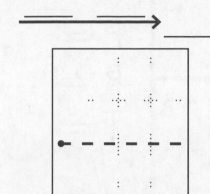 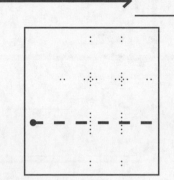 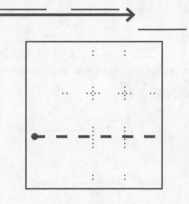

b. _____ – 23 = 67   d. 68 – 23 = _____   f. _____ – 23 = 27

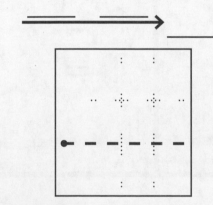 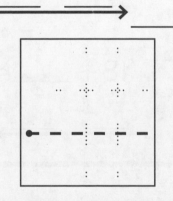 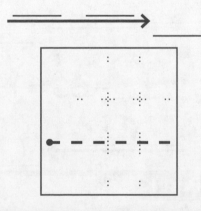

Part E

a. 243 – _____ = 89   c. _____ – 65 = 178

b. _____ + 89 = 154   d. 243 – 178 = _____

Part F

a. 184 – _____ = 76   c. _____ – 184 = 76

b. 184 + _____ = 292   d. 184 + 108 = _____

Copyright © The McGraw-Hill Companies, Inc.

Remedies CONTINUED

Name _____

Part G

H =   H =
P =   P =
R =   R =
T =   T =

Part H

H =   H =
P =   P =
R =   R =
T =   T =

Part I

H =
P =
R =
T =

H =
P =
R =
T =

Part J

a. $79 - \underline{\quad} = 18$

$\Longrightarrow \underline{\quad}$

b. $\underline{\quad} + 31 = 81$

$\Longrightarrow \underline{\quad}$

c. $\underline{\quad} - 63 = 18$

$\Longrightarrow \underline{\quad}$

d. $58 - \underline{\quad} = 17$

$\Longrightarrow \underline{\quad}$

e. $\underline{\quad} - 63 = 32$

$\Longrightarrow \underline{\quad}$

f. $18 + \underline{\quad} = 89$

$\Longrightarrow \underline{\quad}$

Part K

a. $\underline{\quad} - 15 = 65$ b. $65 + \underline{\quad} = 97$ c. $86 - \underline{\quad} = 33$

$\Longrightarrow \underline{\quad}$

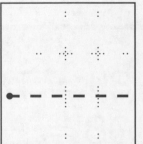

$\Longrightarrow \underline{\quad}$

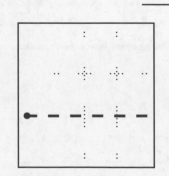

$\Longrightarrow \underline{\quad}$

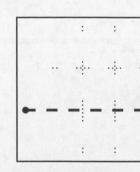

Connecting Math Concepts

Copyright © The McGraw-Hill Companies, Inc.

Remedies CONTINUED

Name _____

Part L

a. _____ + 27 = 68   c. 54 + _____ = 88   e. 88 – _____ = 23

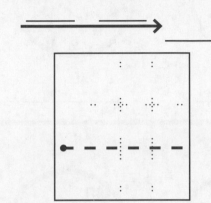   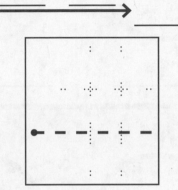   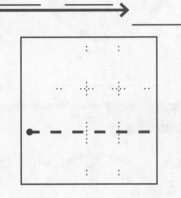

b. _____ – 23 = 67   d. 68 – 23 = _____   f. _____ – 23 = 27

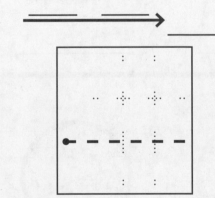   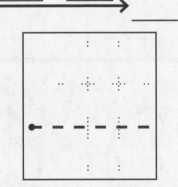   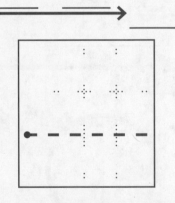

Part M

a.   b.   c.   d.

___:___   ___:___   ___:___   ___:___

Copyright © The McGraw-Hill Companies, Inc.

Remedies CONTINUED

Name _____

Part N

a.

    : _____

b.

    : _____

c.

    : _____

d.

    : _____

Part O

a.

    : _____

b.

    : _____

c.

    : _____

d.

    : _____

Part P

a.

    : _____

b.

    : _____

c.

    : _____

d.

    : _____

Part Q

a.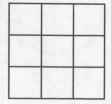

b.

c.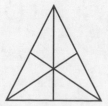

Copyright © The McGraw-Hill Companies, Inc.

Remedies CONTINUED

Name _____

Part R

a.

b.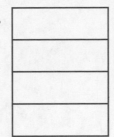

c.

Part S

a.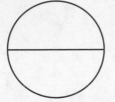

c.

e.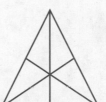

b.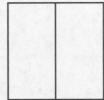

d.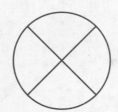

f.

Copyright © The McGraw-Hill Companies, Inc.

Cumulative Test 2

Section A

Part 1

a. _____
b. _____
c. _____
d. _____
e. _____
f. _____
g. _____
h. _____

Part 2

a. $\underrightarrow{5 \quad 1}6$

_____

_____

_____

_____

b. $\underrightarrow{7 \quad 2}9$

_____

_____

_____

_____

Part 3

a. $2 + 7 = 9$ b. $30 + 8 = 38$ c. $15 + 1 = 16$

_____   _____   _____

Part 4

$\underline{90} + \underline{6} = 96$

a. ____ + __ = 37

b. ____ + __ = 81

c. ____ + __ = 90

d. $300 + 40 + 8 =$ ____

e. ____ + ____ + __ = 813

f. ____ + __ + __ = 209

g.
$$\begin{array}{r} 100 \\ 70 \\ +6 \\ \hline \end{array}$$

h.
$$\begin{array}{r} \underline{} \\ + \underline{} \\ \hline 640 \end{array}$$

Copyright © The McGraw-Hill Companies, Inc.

Cumulative Test 2

Name _____

Section A

Part 5

1 + 7	70 + 10	8 + 10
7 − 0	6 − 2	9 + 1 + 6
4 + 1	72 + 10	17 − 10
7 − 1	2 + 6	10 + 5
1 + 8	9 − 7	8 − 3
10 − 10	25 + 10	10 − 6
12 + 1	9 + 2	9 − 3
1 + 5	8 − 2	3 + 5
9 − 8	1 + 7 + 2	6 + 4
18 + 1	13 − 3	9 − 6

Copyright © The McGraw-Hill Companies, Inc.

Connecting Math Concepts

Cumulative Test 2

Section B

Part 6

a.
$$\begin{array}{r} 857 \\ -37 \\ \hline \end{array}$$

b.
$$\begin{array}{r} 386 \\ -381 \\ \hline \end{array}$$

c.
$$\begin{array}{r} 459 \\ -402 \\ \hline \end{array}$$

d.
$$\begin{array}{r} 738 \\ -532 \\ \hline \end{array}$$

Part 7

a.
$$\begin{array}{r} 34 \\ 21 \\ +23 \\ \hline \end{array}$$

d.
$$\begin{array}{r} 72 \\ +18 \\ \hline \end{array}$$

g.
$$\begin{array}{r} 297 \\ 312 \\ +70 \\ \hline \end{array}$$

b.
$$\begin{array}{r} 32 \\ 6 \\ +21 \\ \hline \end{array}$$

e.
$$\begin{array}{r} 56 \\ +14 \\ \hline \end{array}$$

h.
$$\begin{array}{r} 412 \\ 8 \\ +153 \\ \hline \end{array}$$

c.
$$\begin{array}{r} 18 \\ 201 \\ +40 \\ \hline \end{array}$$

f.
$$\begin{array}{r} 8 \\ 52 \\ +37 \\ \hline \end{array}$$

i.
$$\begin{array}{r} 81 \\ 124 \\ +562 \\ \hline \end{array}$$

Part 8

a. _____ b. _____ c. _____ d. _____

e. _____ f. _____ g. _____ h. _____

Part 9

a. _____ b. _____

Copyright © The McGraw-Hill Companies, Inc.

Cumulative Test 2

Section B

Part 10

a. b. c. d.

e. _____ f. _____ g. _____

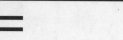

Part 11

= ≠

a. 20 + 35 = 20 + 30 + 4   d. 70 + 11 + 3 = 80 + 3

b. 16 + 1 + 5 = 18 + 5   e. 43 + 9 = 43 + 8 + 1

c. 19 + 10 = 10 + 9 + 10   f. 25 + 40 + 7 = 25 + 49

Part 12

	Blue =	Red =	Yellow =	Green =

	1	2	3	4	5	6
Color						

Part 13

a. 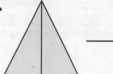 ____

b. 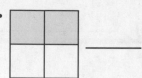 ____

c.  ____

d. 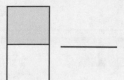 ____

Copyright © The McGraw-Hill Companies, Inc.

Connecting Math Concepts

Cumulative Test 2

Name _____

Section B

Part 14

>   <   =

a.   7   5

b.   8   9

c.   6   3

d.   4   4

e. $20 + 35$ $20 + 30 + 4$

f. $16 + 1 + 5$ $18 + 5$

g. $19 + 10$ $10 + 9 + 10$

h. $70 + 11 + 3$ $80 + 3$

i. $43 + 9$ $43 + 8 + 1$

j. $25 + 40 + 7$ $25 + 49$

Part 15

a. _____ : _____   b. _____ : _____   c. _____ : _____   d. _____ : _____

Part 16

a.

4 feet

10 feet

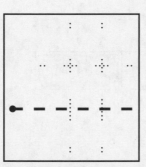

b.

27 feet

16 feet

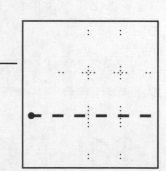

c.

18 feet

8 feet

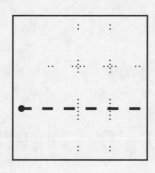

Copyright © The McGraw-Hill Companies, Inc.

Cumulative Test 2

Name _____

Section B

Part 17

1.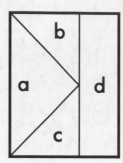

Triangle
Rectangle
Hexagon

▭ = ___   ◁ = ___

▷ = ___   ◣ = ___

2.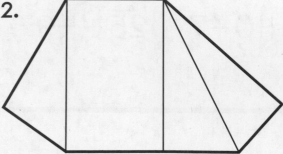

Triangle

Rectangle

Hexagon

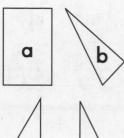

Part 18

a. $20 $10 $5 $5

b. $50 $10 $5 $1 $1

c. $61 + 🪙🪙🪙🪙

d. $33 + 🪙🪙🪙🪙🪙

e. $10 $5 🪙🪙🪙🪙

f. $10 $5 $5 $1 🪙🪙

Connecting Math Concepts

Copyright © The McGraw-Hill Companies, Inc.

Cumulative Test 2

Section B

Part 19

___ + ___ =

a.

___ + ___ = 19

d.

47 + ___ =

b.

___ + ___ = 37

e.

15 + ___ =

c.

___ + ___ = 21

f.

Part 20

a. 24 _____→ 35

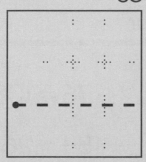

b. 16 423 → ___

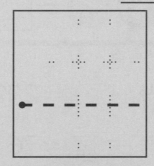

c. ___ 72 → 93

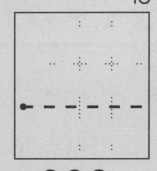

d.
$$\begin{array}{r} 220 \\ -142 \\ \hline 78 \end{array}$$

e.
$$\begin{array}{r} 181 \\ -142 \\ \hline 39 \end{array}$$

f.
$$\begin{array}{r} 220 \\ +181 \\ \hline 401 \end{array}$$

═══→ ___

═══→ ___

═══→ ___

g. ___ − 23 = 65

h. ___ + 54 = 75

i. 89 − ___ = 63

═══→ ___

═══→ ___

═══→ ___

Copyright © The McGraw-Hill Companies, Inc.

Cumulative Test 2

Name _____

Section B

Part 21

a. 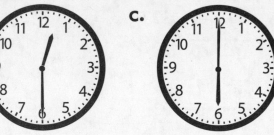 = ⬜ one dollar

b. = ⬜ one dollar

c. = ⬜ one dollar

d. = ⬜ one dollar

Part 22

a. $184 + 108 =$ ____   **c.** $184 +$ ____ $= 292$ **e.** ____ $- 76 = 108$

b. ____ $- 184 = 76$ **d.** $184 -$ ____ $= 76$ **f.** ____ $+ 184 = 260$

Part 23

a. ____ : ____

c. ____ : ____

e. ____ : ____

b. ____ : ____

d. ____ : ____

f. ____ : ____

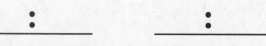

Copyright © The McGraw-Hill Companies, Inc.

Connecting Math Concepts

Cumulative Test 2

Name _____

Section B

Part 24

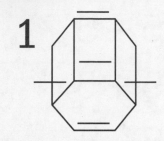 1   2 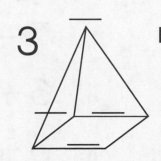   3 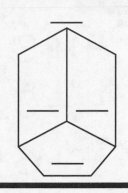   4

Part 25

a. 9 − 3	f. 10 − 2	k. 3 + 5
b. 9 + 2	g. 3 + 6	l. 6 + 4
c. 8 − 5	h. 6 + 6	m. 12 − 10
d. 11 − 5	i. 11 − 6	n. 12 − 6
e. 10 + 2	j. 11 − 9	o. 5 + 6

Part 26

a.

____ + ___ + ___ = ____

b.

____ + ___ + ___ = ____

c.

____ + ___ + ___ = ____

d.

____ + ___ + ___ = ____

Copyright © The McGraw-Hill Companies, Inc.

Cumulative Test 2

Name _____

Section B

Part 27

a. $4 + 3$	f. $7 - 3$	k. $5 - 5$
b. $4 - 3$	g. $6 - 3$	l. $5 + 5$
c. $4 + 6$	h. $6 + 3$	m. $8 - 4$
d. $4 + 4$	i. $10 + 5$	n. $3 + 3$
e. $10 - 4$	j. $10 - 5$	o. $8 - 3$

Part 28

a. $16 - 8$	f. $6 - 3$	k. $11 - 3$
b. $7 + 7$	g. $6 + 6$	l. $7 - 3$
c. $8 - 4$	h. $14 - 7$	m. $7 + 3$
d. $8 + 8$	i. $10 - 7$	n. $11 - 8$
e. $10 - 5$	j. $8 + 3$	o. $10 - 3$

Connecting Math Concepts

Copyright © The McGraw-Hill Companies, Inc.